Nella Burnett-Stuart
Liz Taylor

ZED
The Magician

U0061905

商務印書館

CONTENTS

出版說明	1
使用說明	2
ZED *The Magician*	3-24

ACTIVITIES

WWF Form	25
Endangered Animals	26-27
Bingo!	28-29
Treasure Island	30-31
Crack the Code!	32
Key	33-34
Glossary	35

Originally published by Black Cat Publishing under the title:
Zed The Magician
© 2001 Black Cat Publishing
An imprint of Cideb Editrice, Genoa, Canterbury

The copyright of this Chinese edition is owned by
The Commercial Press (H.K.) Ltd.

Name of Book: Zed The Magician
Author: Nella Burnett-Stuart, Liz Taylor
Editors: Claudia Fiocco, Kathryn Pass
Design and Art Direction: Nadia Maestri
Illustrations: Laura Scarpa
Layout: Simona Corniola

系 列 名： Quality English Learning for Kids · II
書　　名： Zed The Magician
責任編輯： 黃淑嫻
出　　版： 商務印書館 (香港) 有限公司
　　　　　香港筲箕灣耀興道 3 號東滙廣場 8 樓
　　　　　http://www.commercialpress.com.hk
印　　刷： 美雅印刷製本有限公司
　　　　　九龍觀塘榮業街 6 號海濱工業大廈 4 樓 A
版　　次： 2008 年 7 月第 2 次印刷
　　　　　© 商務印書館 (香港) 有限公司
　　　　　ISBN 13-978 962 07 1710 9
　　　　　ISBN 10-962 07 1710 4
　　　　　Printed in Hong Kong

出版說明

　　學英語當然要學優質的，有品質才能讓人有信心。我們一直積極提倡學習優質英語的理念，並且為學習者提供過多元化的優質英語材料，像《Black Cat 優質英語階梯閱讀》就十分成功，至今已出版近 60 本。鑑於良好的英語能力最好從小培養，我們於是出版這一套適合七至十歲兒童的優質英語閱讀讀本 "Quality English Learning for Kids"。

　　培養兒童對英語的興趣，須從趣味和簡易兩方面入手。圖文並茂，聲文結合這兩大特點對學習英語甚有幫助。"Quality English Learning for Kids" 承續本館出版優質英語書的理念，全書彩圖精美，附 CD 朗讀內容及聆聽練習，形式多元化，有故事讀本（story books）、圖畫讀本（picture readers）、戲劇讀本（drama readers）及互動讀物（interactive readers）四大類，提供不同的學習功能。故事讀本和圖畫讀本可供兒童看圖講故事；戲劇讀本完全用對白編寫，培養脫口而出講英語的習慣，適合家庭裏作簡單的角色扮演，或者小學生在課堂作簡單的演出。

　　針對兒童學習英語的需要，本系列提示家長為兒童設定學習目標，並且說明如何達標，另備生詞表和語法知識點，讓兒童在家長協助下掌握生詞用法，認識簡單的句子結構和了解語法要點。

　　"Quality English Learning for Kids" 吸引兒童對閱讀產生興趣，逐步引導他們參與愉快的閱讀旅程。在這個旅程中，家長是重要的導航者，透過對兒童的悉心鼓勵，循循善誘，進一步加強親子關係。

商務印書館
編輯部

使用説明

 如何使用本書？

本書為互動讀本（interactive reader），適合課堂使用或親子共讀，每頁均圖文並茂，除正文外，還加上了與該頁內容有關的話題，供家長或老師引發小孩子的興趣，讓小孩子就話題內容提出個人的想法，發揮想像力和培養創意思維。

本故事的正文由敘述文字和人物對話組成，兩者相輔相成。敘述文字讓小孩子掌握第三身複述句法，人物對話則讓小孩子掌握第一身直述句法及對答技巧。本書附有 CD，小孩子可邊聽邊讀，提高英語聽説能力。

 本書的學習目標是甚麼？

教師或家長可為孩子定出以下學習目標。

使用本書後，孩子必須學會：

(a) 用現在進行時描述自己正在做甚麼事（say what I am doing）

(b) 説出自己喜愛的動物和音樂（say what animals/music I love）

(c) 談天氣狀況（say what the weather is like）

(d) 描述某人的特徵（describe someone）

(e) 一面聆聽，一面填表格（listen and complete a form）

 本書有哪些重點生詞和語法學習項？

(a) 重點生詞：本書的重點生詞分五大類，包括動物（animals）、天氣和季節（weather and seasons）、時間（the time）、日期（days of the week）及魔術（magic）。

(b) 語法學習項：

　i) 句子結構：主要為基本句型"主語 + 動詞 + 賓語"，也運用了大量的前置詞（prepositions）和副詞（adverbs）組成較長的句子。

　　例子：• "Suddenly, they see a police car outside the school."（頁 7）

　　　　　• "They leave the envelope at the bottom of the statue."（頁 14）

　　　　　• "They meet in front of the theatre on Sunday afternoon."（頁 16）

　　　　　• "On Monday morning the children are sitting in their classroom."（頁 23）

　ii) 語法要點：書內的常用語法包括了

　　• present continuous（例如頁 3，"Martin's father is reading the newspaper."）

　　• prepositions　　　（例如頁 19，"They are behind the rubbish container."）

　　• imperatives　　　（例如頁 10，"Listen to the message."）

　　• time expressions　（例如頁 9，"It's 4 o'clock. It's time to go home!"）

　　　　　　　　　　　（例如頁 11，"It's Saturday...It's a cold and windy day."）

　　　　　　　　　　　（例如頁 16，"They meet in front of the theatre on Sunday afternoon."）

　　• wh- questions　　（例如頁 6，"Where's Bonnie?"）

ZED
The Magician

Martin's father is reading the newspaper.

1. ANOTHER BIG ROBBERY!

2. THAT'S FOUR IN THIS AREA!

What are **you** doing now?

3. I HOPE THEY CATCH THE ROBBERS SOON.

Martin is a member of the WWF.
He loves animals.
He and his friends are organising a school party –
a special party! A party with a magic show.

It's 5 o'clock.
Zed the Magician is on stage.
He is starting his magic show.

Zed waves his magic wand. ABRACADABRA...

... AND NOW FOR MY FINAL TRICK!

He opens the box. It's empty!

WHERE'S BONNIE? BONNIE ISN'T IN THE BOX! WOW ... MAGIC!

It's 8.40 am.
Martin and Tony are walking to school.
Suddenly, they see a police car outside the school.
A policewoman is going into the school.

The headteacher is explaining the problem.
He isn't very happy.
'There was a robbery in the school yesterday.
All the money for the *Whales in Danger* is gone!
... and the computers are gone!'

It's 4 o'clock. It's time to go home!
Martin and Tony are ready.
They are standing near the door.
Martin sees a small red phone in the corner. He takes the phone and puts it into his pocket.
He wants to take it to the police station.

1. WHAT'S THAT UNDER THE TABLE?

2. IT'S A PHONE.

What do **you** do at 4 o'clock?

3. HEY... DO YOU THINK IT'S THE ROBBERS' PHONE?

4. MAYBE.

9

The children are waiting for the bus. They are looking at the phone. Tony presses a button.
They listen to a message.

It's Saturday. Martin, Tony and Annie meet at the lake. It's a cold and windy day.
All the boats on the lake are empty.
Except one...

The man is waiting next to the statue
of Queen Victoria.
He is angry. He puts an envelope at the bottom
of the statue.
The children are watching him.

The children get into a boat.
They row to the statue.
Annie quickly takes the envelope.

Inside the envelope there is a message.
The children read the message.

2.PERHAPS THEY WANT TO ROB A PUB...

3.... OR A THEATRE!

4.... OR A CINEMA!

What are you reading?

5.COME ON! LET'S GO! LEAVE THE ENVELOPE!

1. THE MONEY IS IN THE BOX.
THE FIRE EXIT IS ALWAYS OPEN.
LOOK FOR A BLUE CAR.

They leave the envelope at the bottom of the statue.

14

It's 4.30 pm on Saturday.
The children are outside a bookshop.
They love books.
They see a poster in the shop window.

Martin, Annie and Tony want to catch the robbers. They meet in front of the theatre on Sunday afternoon. It's 2.00 pm.

Is it raining where **you** are today?

1. YOU BOYS HIDE OUTSIDE NEAR THE FIRE EXIT AND WAIT FOR THE ROBBERS.

2. BUT IT'S RAINING...

3. WHERE ARE YOU GOING?

4. I'M GOING INSIDE!

16

Annie is inside the theatre.
She sees Zed and Bonnie.
They are talking. Annie stops and listens.

Annie is sitting in the theatre.

The spotlight is on Zed.

Bonnie is stepping into a long box.

NOW Annie understands!

Martin and Tony are hungry, bored and wet.
They are behind the rubbish container.
Suddenly they see Annie.
She is at the fire exit.

Annie is looking out of the window.
She sees a blue car.
The man in the car has red hair!
It's the man from the lake!

She locks  the fire exit door.

Zed is waving his magic wand. ABRACADABRA!
He opens the box. It's empty! But ... where's Bonnie?

Bonnie runs to the fire exit.
She is pushing the door.
She can't open the door... it's locked.
Martin and Tony hold her.
Then the police arrive.
They arrest Bonnie, Zed and the man with the red hair.

On Monday morning the children are sitting in their classroom. The teacher is talking about the robbery.

Martin is lying on his bed.
He is thinking about the whales.
Wow! All that money just for them!

WWF (World Wildlife Fund) FORM

 Listen and complete the form.

(🐼) **WWF Form**

Name : _Tim_
Age : ___
Birthday : ___
New Member : ☐ YES   ☐ NO
Address : _Albany Road_
_____ _W2_
Subscription Cost : £ ___
Free Poster : ___

- -

This is your personal WWF Form. Complete it now.

(🐼) **WWF Form**

Name : ___
Age : ___
Birthday : ___
New Member : ☐ YES   ☐ NO
Address : ___
_____
Subscription Cost : £ ___
Free Poster : ___

ENDANGERED ANIMALS

Look at the pictures of these animals. Read the descriptions. Then, match them to the right picture.

1 It's very big and long. It lives in the sea but it can't breathe under water. It eats small fish. It's a

....

Mountain Gorilla

2 It lives in the jungle. It eats meat and hunts at night. It's got black and orange stripes. It can swim and jump. It's a ....

Nile Crocodile

Rhinoceros

3 It doesn't eat meat. It has very long arms and long hair. It is usually black. It can climb trees.
It's a ....

4 It eats plants and grass all day long. It's very heavy. It likes mud. It has horns on its nose.
It's a ....

Blue Whale

5 It lives in the water. It eats fish and birds and sometimes humans! It has big sharp teeth and a long strong tail.
It's a ....

Tiger

BINGO!

1 **Listen to the words and the sounds, and point to the pictures as you hear them.**

2 **Now listen to the sounds and point to the right picture. Tick the box ✓ if you like it. Cross the box ✗ if you don't like it.**

c. Drill ☐

b. Sparrow ☐

a. Monkey ☐

d. Motorbike ☐

e. Trumpet ☐

g. Telephone ☐

f. Lion ☐

i. Cat ☐

k. Elephant ☐

h. Clock ☐

j. Blue Whale ☐

28

3 **Now let's play BINGO!**

   a **Choose 6 things.**

   b **Write one name in each square.**

   c **Now listen to the recording and play BINGO!**

1

~~Monkey~~	~~Telephone~~	~~Motorbike~~
~~Blue Whale~~	~~Cat~~	~~Clock~~

Bingo!

2

3

4

5

To play Bingo again with your friends, write the name of each sound on a flash-card and continue to call out the sounds!

29

(8) TREASURE ISLAND

Can you find: a parrot, a monkey, a sparrow, an elephant, a crocodile?
Match the correct word to the pictures on the map.
Write the number in the circle.

1 river
2 mountains
3 lake
4 forest
5 beach
6 boat
7 palm trees
8 village
9 cave

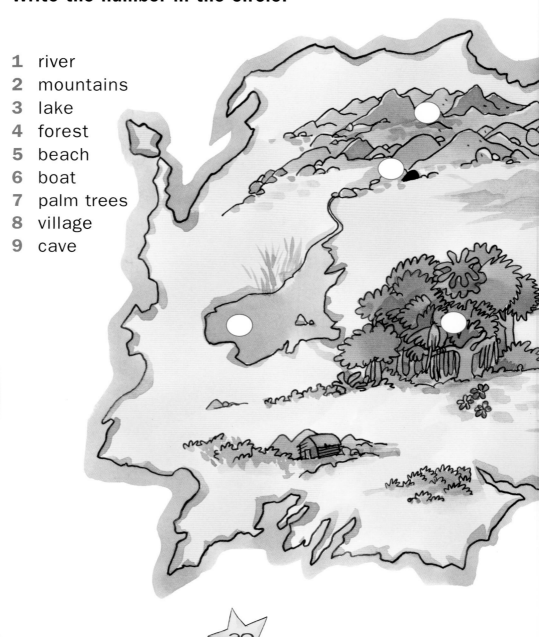

CRACK THE CODE

A =   E =   I =

O =   U =

First crack the code.

Gₒ nₒrth tₒ thₑ mₒᵤntᵢns ₐnd wₐlk thrₒᵤgh thₑ cₐvₑ

Gₒ sₒᵤth ₐnd swᵢm ₐcrₒss thₑ lₐkₑ

Thₑn gₒ sₒᵤth ₐgₐᵢn ₐnd wₐlk pₐst thₑ fₒrₑst tₒ thₑ rᵢvₑr.

Tₐkₑ ₐ bₒₐt ₐnd gₒ ᵤp thₑ rᵢvₑr.

Thₑ whₐlₑ ᵢs ₒn thₑ bₑₐch ᵤndₑr thₑ pₐlm trₑₑs.

Well done! Now look at the map on pages 30 and 31 and find the whale. It is hiding somewhere!

ZED The Magician

Page 25 – WWF Form

Tapescript

WWF secretary: Hello, this is the World Wildlife Fund, Monica speaking. Can I help you?

TIM: Yes I'd like to join the WWF.

WWF secretary: Have you been a member before?

*TIM: No, this is **my first time**.*

*WWF secretary: Membership is **£25 a year**. You get a magazine every month, and a **free animal poster**. Could you give me your name and address?*

*TIM: **Tim Stuart**, and I live at **42 Albany Road, London W2**.*

WWF secretary: Is that Stuart with a W?

TIM: No, with a U. S-T-U-A-R-T.

WWF secretary: And when's your birthday Tim?

*TIM: I was born in **1989**.*

WWF secretary: Tim, could you tell me the day and the month of the year?

*TIM: Oh yes, **January the 21st**.*

WWF secretary: Oh, how interesting you were born on the 21st of January, the same as my youngest brother, who's also called Tim!

WWF Form	
Name :	Tim Stuart
Age :	15 (current year = 2004)
Birthday :	January the 21st, 1989
New Member :	~~YES~~ NO
Address :	42 Albany Road
	London W2
Subscription Cost :	£ 25 a year
Free Poster :	animal poster

Page 26 – Endangered Animals

1 Blue Whale
2 Tiger
3 Mountain Gorilla
4 Rhinoceros
5 Nile Crocodile

Page 28 – Bingo!

Listening 1: open activity
Listening 2: 1 h 2 i 3 g 4 c 5 b
6 f 7 j 8 e 9 k 10 a 11 d

Page 29 – Listening 3

Open activity.

Page 32 – Crack the Code!

Go north to the mountains and walk through the cave.
Go south and swim across the lake.

Then go south again and walk past the forest to the river.
Take a boat and go up the river.
The whale is on the beach under the palm trees.

Page 30 – Treasure Island

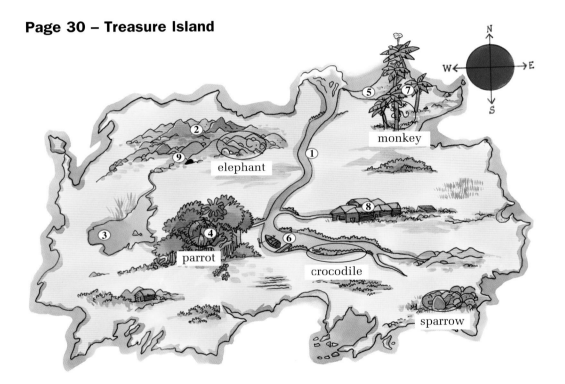

Glossary 生詞表

abracadabra（魔術師表演魔術時的）唸唸有詞

arrest 拘捕

assistant 助手

disappears 消失

except 除了…之外

explaining 解釋

final 最後的

instructions 指示

locked 上了鎖的

magic wand 魔術棒

magician 魔術師

member 成員

message 訊息

on stage 在台上演出

organizing 籌備

presses 按

pub 酒吧

robbery 劫案

special 特別的

spotlight（聚光燈的）光

statue 塑像

theatre 劇院

trick 把戲

waves 揮動

whales 鯨